SCIENCE IN ACTION

EXPERIMENTS IN PHYSICS

The books in the SCIENCE IN ACTION series are

Light and Sound

The Living World

The World of Numbers

Fun with Chemistry

Projects in Physics

Experiments in Physics

SCIENCE IN ACTION

The Marshall Cavendish Guide to Projects and Experiments

Compiled and revised by
Sue Lyon

EXPERIMENTS IN PHYSICS

Projects created by
Paul Berman and Keith Wicks

PALM BEACH COUNTY
LIBRARY SYSTEM
3650 Summit Boulevard
West Palm Beach, FL 33406-4198

Marshall Cavendish Corporation · New York · London · Toronto · Sydney

Editorial Staff

Series Editor — Sue Lyon

Assistant Editors — Nigel Rodgers

Caroline Macy

[VOLUME 6]
Art Editor — Kay Carroll

Production Manager — Carol Milligan

Managing Editor — Alan Ross

Editorial Director — Maggi McCormick

Publishing Coordinator — Robert Paulley

This Edition published 1993

Published by Marshall Cavendish Corporation
2415 Jerusalem Avenue
North Bellmore
New York 11710

Typeset by Quadraset Ltd.
Printed in the USA by Worzalla Publishing Company, Wisconsin.

Science in Action derived from material previously published as Volume 25, Growing Up With Science, by H.S. Stuttman Inc.

© Marshall Cavendish Limited
MCMLXXXVII, MCMLXXXVIII

Library of Congress Cataloging-in-Publication Data available upon request
ISBN 0-86307-020-5(set)
ISBN 0-86307-942-3(vol)

Contents

Projects marked ⟊ need adult supervision.

Introduction

This book will help you learn more about science and technology. It includes experiments, projects, puzzles and even some tricks. Some of the experiments and projects are very easy. Others are a little harder (they are marked ✛) and you will need help from one of your teachers or parents. You don't have to begin on page 8—look through the book and start with something you like—but remember that good scientists
- make a record of their work
- have a clean and tidy laboratory
- most important, keep themselves safe (always read pages 40 to 42 before you begin).

SIMPLE EXPERIMENTS

Many of the everyday appliances and tools that we take for granted owe their existence to the work of physicists of the past who, in their experiments, discovered some of the fundamental physical laws of the universe.
In the following pages, there are simple experiments that demonstrate some of these laws: see the effects of inertia, friction and surface tension; discover some of the properties of liquids; and make your own clouds in a bottle.

Perpetual motion

Once started, a perpetual motion machine will run forever without needing fuel or any other energy supply. Over the centuries, many attempts at such machines have been made. In the one shown here, marbles roll along tubes, which are fixed to a disk. The idea is that the marbles will move so that their weight keeps the disk spinning in one direction.

Procedure
You will need **two large marbles** and two **identical cardboard tubes** in which they can roll.
1. Cut a disk from a **sheet of stiff cardboard**. The diameter of the disk should be slightly greater than the length of the tubes.
2. Put one marble in each tube. Seal the ends.
3. Push a large round **nail** through the center of the large disk. Then, using a **thick needle**, stitch the tubes to the disk with **strong thread**. The nail should protrude from the other side.
4. Have an assistant hold the nail level while you spin the disk. As the disk turns, the marbles will roll along the tubes, one after the other.

But do not be too disappointed if the disk soon stops spinning! As the marble on the left rises, its weight opposes that of the falling marble on the right. So there is no lasting force to overcome the friction between the disk and the nail on which it spins.

In fact, friction has defeated every attempt in history to build a perpetual motion machine.

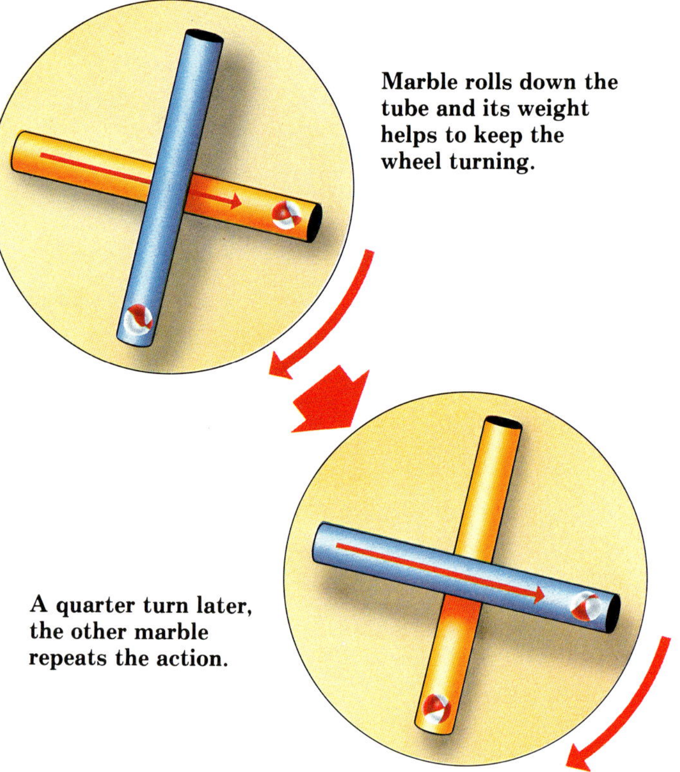

Marble rolls down the tube and its weight helps to keep the wheel turning.

A quarter turn later, the other marble repeats the action.

Wheels in motion ⊹

When a wheel spins quickly, it tends to resist any attempt to change the direction in which its axis points. This is the principle of the gyroscope and gyrocompass. It is also the reason why a bicycle is easier to keep upright when it is moving at speed. However, if you *do* force the axis of a spinning wheel to point in a different direction, a strange effect occurs.

Procedure

1. Use a **bicycle wheel** and support the spindle horizontally using **two small cardboard tubes** held in the hands. Then ask someone to give the wheel a spin (figure 1).

2. Try to keep the wheel axis horizontal while you turn sideways in the direction shown by the arrow. Instead of moving in the direction you want, the wheel will appear to have a mind of its own.

3. Observe which way the wheel tends to move. It should try to flip over to make its axis vertical (figure 2). This strange and powerful effect is called precession.

Did you know . . .

Wheels are not only vital to transportation, they can also be used to store energy in the shape of flywheels. These are huge wheels with their weight concentrated on their outer rims. Once set in motion, they spin almost for ever.

You will need—

1 bicycle wheel
2 small cardboard tubes

Take care!

Hold the cardboard tubes so that your hands are away from the spinning bicycle wheel.

Inertia

Inertia is the inherent or "built-in" tendency of everything to stay as it is without changing. Galileo Galilei (1564–1642), the Italian astronomer and physicist, first put forward the idea of inertia. He imagined a perfectly smooth sphere on a perfectly smooth slope and he said that if you could do away with friction the sphere would roll increasingly quickly down the slope. To make it go up the slope, it would have to be pushed or have a "force" applied to it. Indeed, even to make it stay still it would have to be held in place by force. Galileo suggested that if you had no friction and an endless level surface, a sphere set rolling on it would never stop of its own accord.

In the ordinary way, friction, which is the rubbing force that happens between any surfaces that are touching each other, does affect what happens and slows things down. If it did not, perpetual motion machines, which have been dreamed about for centuries, might work. As it is, even the air about us slows things down. Modern car designers spend a lot of time and thought trying to make air slip over their cars with less and less friction. This is important because a car is a large, heavy object and overcoming its inertia and making it start and stop needs a lot of force. This force is supplied by the engine or the brakes. Speeding up and continuing to move takes a lot of energy which the engine produces by burning gasoline. The sleeker the car is, the less friction there is between the car's body and the air, and the less energy it takes to move it.

Demonstrating inertia

The laws of inertia also enable magicians to perform the amazing trick of removing a tablecloth out from under the dishes. The cloth is moved so quickly and cleanly that there is not enough force to overcome the inertia of the cups and saucers on the cloth and so they do not move. You should not try this trick yourself as the dishes might get broken, but there are some things you can do, although you may need to practice. For one experiment with inertia, you will need to build a pile of six **checkers**—make sure the surface they are on is a slippery one, not carpet or cloth. Now take a **ruler**, lay it flat on the surface and QUICKLY slide it across and under the column, knocking the lowest checker away. The other five should stay in place. An easier version of the tablecloth trick is to place a sheet of smooth, **strong paper** half on and half off a table edge with an unbreakable object standing on it. Pull the paper quickly enough and the object will stay standing just where it is. Another impressive demonstration of inertia is to get a tall piece of **wood** with a flat bottom—the piece should be about ten times as high as it is wide but you may want to practice with a piece five times as high. Cut a strip of **paper** the same width as the wood and stand the wood on the paper near the edge of a smooth tabletop. Hold the paper taut with one hand and hit the paper hard with a **ruler**. This should remove the paper from under the wood while leaving the wood standing.

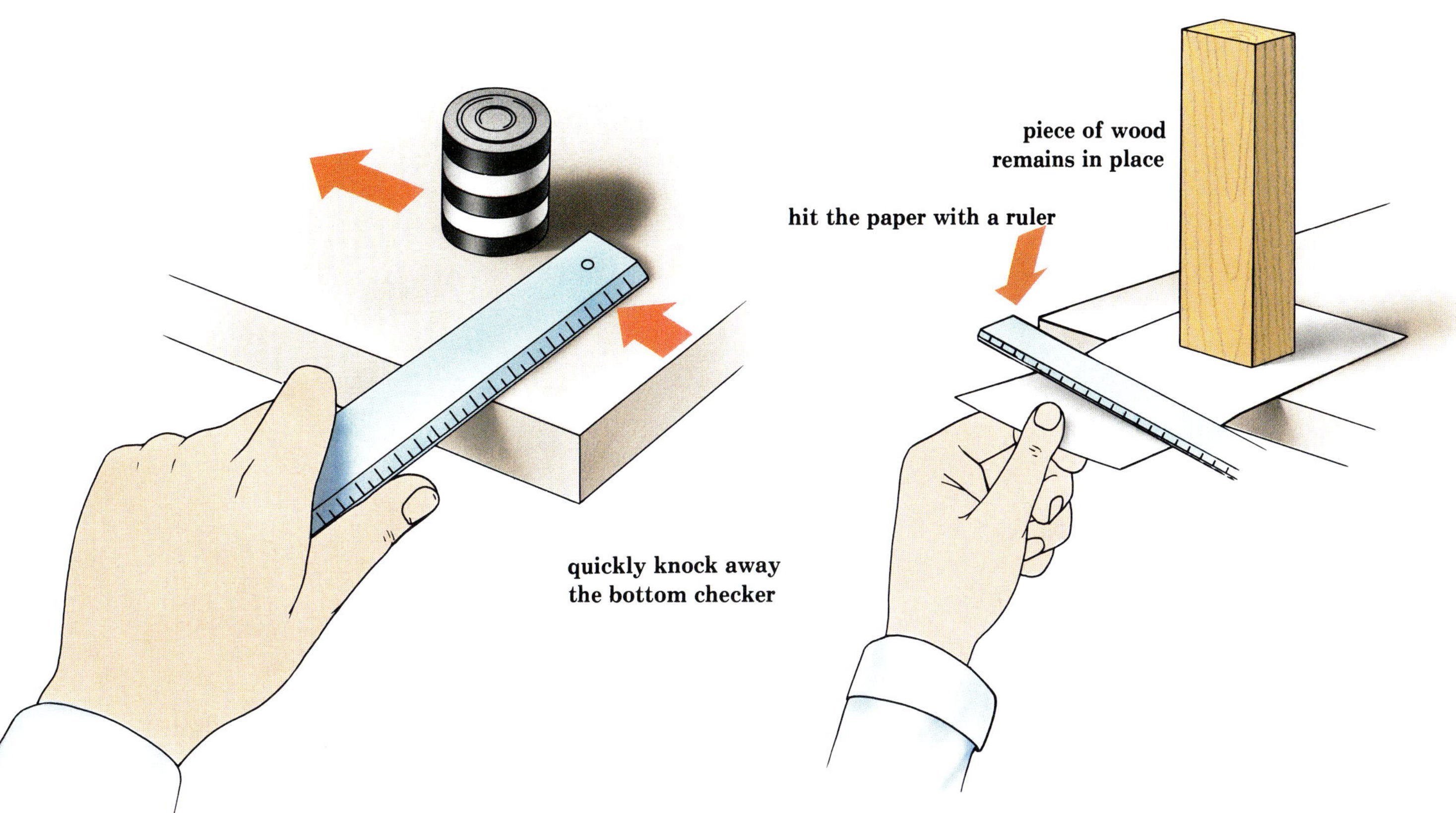

quickly knock away
the bottom checker
hit the paper with a ruler
piece of wood
remains in place

Magnetic compass

The earth acts as a giant magnet, so it has an effect on other magnets. See how the earth's magnetism makes a hanging bar magnet point in one particular direction. And use this principle to construct a magnetic compass.

Theory of magnetism

The poles of a magnet are the parts that have the strongest magnetic effect. The ends of a typical bar magnet, for example, will readily attract small pieces of iron or steel. But little or no attraction will be detected near the middle of the bar. This shows that the poles are at the ends of the magnet.

The poles of a magnet are usually referred to as *north* and *south*. Their full names are the *north-seeking* and *south-seeking* poles. The north-seeking pole is so called because, when the magnet is free to move, it will turn to point toward the earth's North Pole. This effect may seem strange because it appears to contradict the fundamental laws of magnetism. These state that *like* poles repel and *unlike* poles attract each other. If, for example, you bring the north pole of one magnet close to the north pole of another magnet, you will feel a force trying to keep them apart. So why should the north pole of a magnet swing toward the earth's North Pole? The answer is that the region known as the North Pole actually contains a *south* magnetic pole. And the *north* magnetic pole of the earth is in the region called the South Pole.

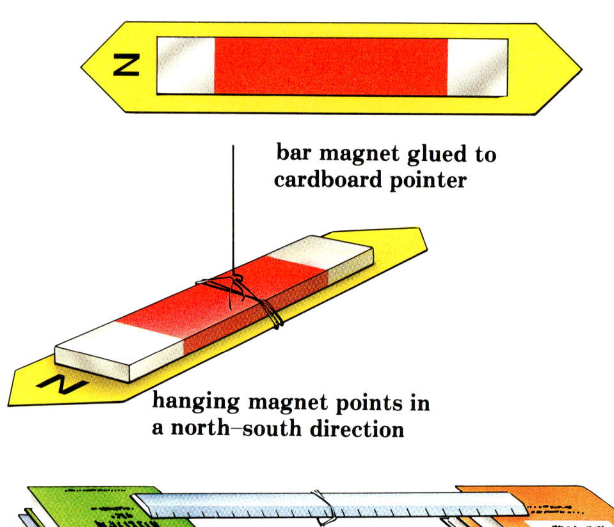

bar magnet glued to
cardboard pointer

hanging magnet points in
a north–south direction

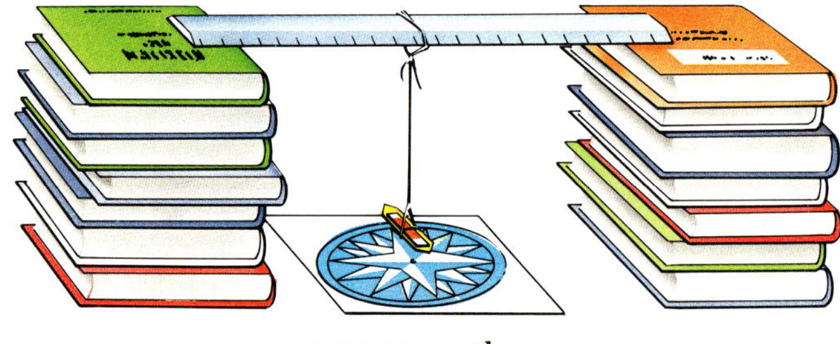

compass card

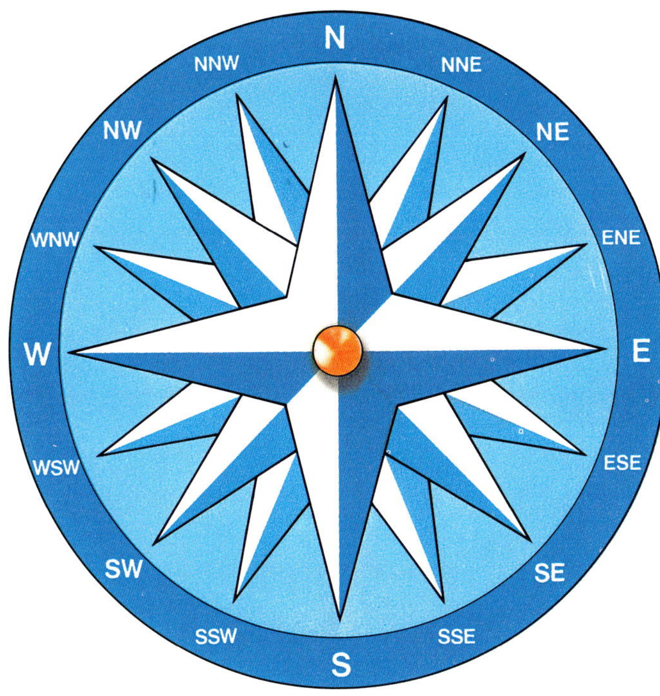

the compass card

The magnetic compass

A typical magnetic compass has a magnetized needle which is mounted on a pivot. The attraction of the earth's magnetic poles makes the needle turn on the pivot until it points in a north–south direction.

In the compass described below, construction is simplified by using a bar magnet instead of a magnetic needle. And the magnet is suspended from a thread instead of being mounted on a pivot.

Construction

1. Tie a piece of **thread** to the middle of a **bar magnet**. Then make a **cardboard pointer** and **glue** the magnet to it.
2. Suspend the magnet just above the surface of a **table**. A convenient support can be made using two piles of **books** bridged by a **wooden rod**.
3. Let the magnet come to rest and note the position of the pointer. It should now lie in a north–south direction. Mark the end pointing north with the letter N. If you are not sure which is north and which is south, observe the position of the sun around midday. In the United States and other countries north of the equator, this direction is approximately south.
4. Make a compass card and pin it directly below the pointer. Turn the card so that its north point lines up with the N marked on the pointer. Any required direction can now be read from the card.

Friction

1 empty plastic carton
strong nylon thread
sand
4 pencils
1 ruler

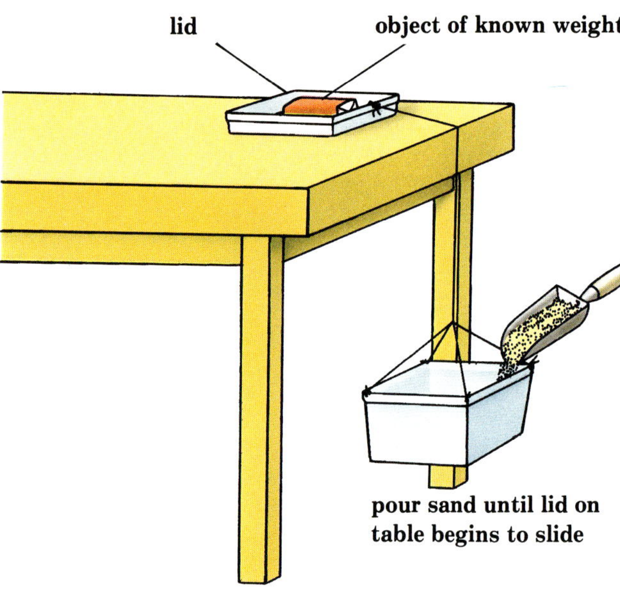

When viewed through a powerful microscope, even the smoothest of surfaces appears as a mass of peaks and valleys. When two surfaces are placed together, some of the peaks of each will drop into the valleys of the other. If one surface is slid over the other, these peaks will be bent and broken as the surfaces drag across each other. This may lead to scratching and wear. Normally, wear is unwanted because it reduces the life of moving parts but a product such as sandpaper, for instance, is meant to cause wear. The heat generated in some materials as they slide over one another is enough to melt small areas, forming microwelds which make the surfaces stick together. This heating effect is also normally unwanted, but some machining processes use it to perform permanent welding operations, and it is what lights a match.

The overall effect of this resistance to sliding is called friction. Friction is sometimes useful; automobile braking and clutch systems rely on it, walking would be impossible without it and even nails would be useless without it. Often friction is unwanted and, unless reduced by lubrication, can lead to moving parts wearing out or overheating. Lubrication works by filling the valleys of each surface with a fluid, therefore preventing the peaks of the other surface from becoming caught in the valleys.

Normally the frictional force which must be overcome to start a surface sliding over another (static friction) is greater than that required to keep it moving (dynamic friction). This is because the two surfaces settle against each other and interlock more tightly when at rest. The illustration shows how to make a simple rig that will let you study both static and dynamic friction.

Procedure

1. Place the top from an empty plastic **carton** (such as an ice-cream carton) on a smooth, flat **table top**.
2. Cut a piece of strong **nylon thread** slightly longer than the height of the table top from the ground.
3. Tie one end of this thread to a hole cut in the edge of the plastic carton lid.
4. Make holes in the corners of the carton, near the top.
5. Tie lengths of thread between the holes to make a handle for the carton.
6. Tie the free end of the thread from the lid to the middle of the handle.

To use the rig, place an object of known weight, such as a candy bar, on the lid. Move the lid until the carton hangs on its thread. If the lid slides when released, you need to put a heavier weight on it. When you have put enough weight on the lid to prevent it from sliding, you can measure the static frictional force by slowly pouring sand into the carton until the lid starts to slide. The weight of the carton plus sand is then equal to the static frictional force. If you now want to find the dynamic frictional force, remove some sand, a little at a time, until the lid does not quite continue moving after you push it gently toward the table edge. The weight of sand then in the carton is equal to the dynamic frictional force.

Try seeing how the frictional forces vary as you add extra weights to the lid, or cover the base of the lid with different materials, such as cardboard or cloth, or lubricate the base with oil, plain water or soapy water (protecting the table surface first if necessary).

When you have finished these experiments, you may be able to answer this problem:

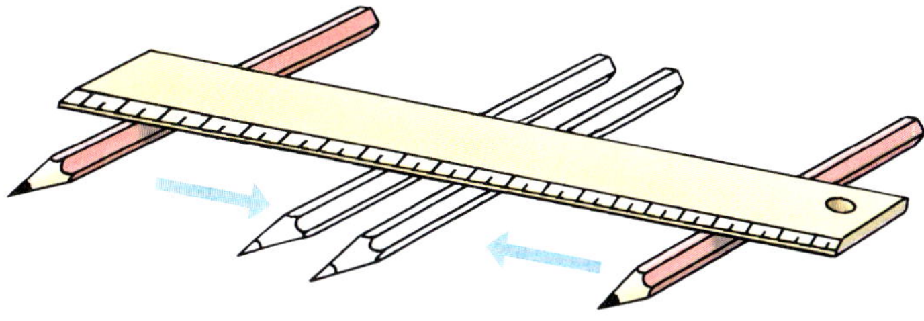

The Ruler Puzzle

If a ruler is laid across two pencils on a table and the two pencils are slowly moved together, the ruler will alternately slide against first one and then the other so that the pencils end up at the midpoint of the ruler. If the pencils are now moved apart, the ruler will slide against only one pencil until it reaches the end, while the other pencil remains in the middle. Why is this?

Greenhouse

Even without an internal heater, a greenhouse can be used to germinate seeds or to grow plants that would not survive out of doors. Greenhouses are made of glass, which does not readily retain heat, so how does a greenhouse work? Find out with this simple experiment, which must be done on a sunny day.

Procedure

1. Fill two **plastic dessert pots** with **water** and place them next to each other outside in the sun. Put a **glass bowl** upside down over one of the pots.

2. Leave the pots in the sun for about an hour. Then remove the bowl and feel the water in the two pots. You should find that the water in the pot that was covered by the bowl is warmer than that in the uncovered pot.

3. This happens because the heat rays of the sun pass easily through the glass and heat the water in the pot, and in a greenhouse cause the water exhaled by the plants to condense into water vapor. The water vapor gives off heat, but at a lower temperature than that of the sun, and so the heat cannot pass through the glass and the space inside the greenhouse heats up.

Surface tension

Throughout a liquid, the molecules are attracted to one another. At the surface of the liquid, the forces of attraction lead to an effect called surface tension. The two soap-film tricks shown here demonstrate surface tension in action.

The jumping wire

1. Put some **liquid soap** in a **glass** and add **water** to make a strong, soapy solution.
2. Bend a piece of **wire** to form a rectangular frame and handle, as shown above. The frame should be small enough to fit into the glass.
3. Dip the frame into the solution and then remove it. The frame should now be covered with a soap film.
4. Hold the frame horizontally and place a straight piece of wire across it.
5. Break the soap film on one side of the wire with your finger. Forces of surface tension on the other side will pull on the wire, causing it to jump from the frame.

The magic loop

1. Tie a short piece of **thread** to form a loop.
2. Form a soap film on a wire frame, as previously described. Carefully place the thread on the film.
3. Now break the film inside the loop. Surface tension forces around the loop will pull it into a neat circle.

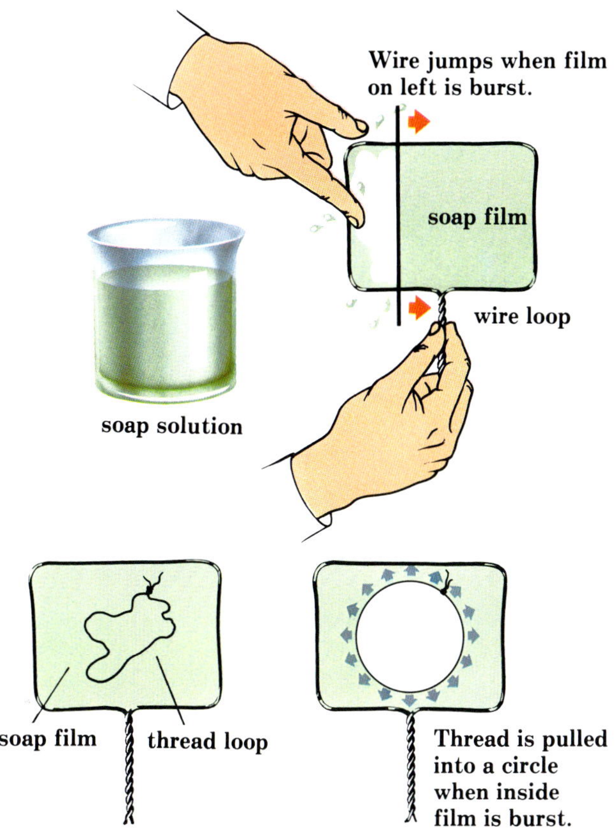

You will need–

liquid soap
1 glass
2 pieces of wire
thread

Water

You will need—

1 funnel
sand and gravel
1 empty glass jar
soil
water

Water is one of the most common substances on earth. It is also the most important for the survival of the human race and all other living things. A person can live for several weeks without food, but without water would die within a few days. Water is a chemical mixture of hydrogen and oxygen, both of which are gases at ordinary temperatures and pressures. Combined together they make water, which is liquid at normal temperature and pressure and does not turn into a gas until its temperature is raised to 212° Fahrenheit (100° Celsius)—called the boiling point. Water turns to ice and becomes solid at 32° Fahrenheit (0° Celsius). Seventy percent of the earth's surface is covered with water. There are the vast oceans, the biggest of which are the Atlantic and the Pacific. There are also the Polar ice caps and the snow-covered tops of high mountain ranges.

All sources of water are part of the water cycle. Water vapor, evaporated by the sun, rises into the air where it cools and forms clouds, eventually falling back onto the earth either as rain, snow or hail. It then seeps back into the rivers and oceans and the cycle starts again.

The power of water

Although it is so essential to life, water can also be very destructive. When a river or an ocean floods the land, lives are often lost and buildings destroyed. The oceans and rivers are also changing the shape of our country as they move sand and earth around, taking them from one place—this is called erosion—and leaving them in another—this is called deposition. Rivers can cut huge ravines by wearing away the gravel they flow over with the constant rubbing of pebbles and silt. Probably the most spectacular example of this in the world is the Grand Canyon in Arizona where the Colorado River has slowly cut a channel for itself and now flows hundreds of feet below its original level.

Filtering water

Pure water is needed in great quantities by modern-day households who use it and then throw it away. The used water is often dirty and if it were just put right back into a river or the ocean it could pollute clean water. In order to prevent this, all the dirty water (called effluent) is collected by underground pipes—called sewers—and taken to special treatment areas where it is passed through filter beds to clean out all the impurities. Filter beds are made up of thick layers of gravel and sand. You can see how this works at home very easily. You will need a **funnel** (the larger the better), some clean **sand** and **gravel**, an empty **glass jar**, and **soil** and **water** mixed together.

All you need to do is to put separate layers of sand and gravel into the funnel, starting with larger pieces of

gravel at the bottom and finishing with sand. Now stand the funnel in the empty jar and slowly trickle the dirty water over the surface of the sand. You will see that the water that falls into the jar underneath is nice and clean.

Did you know . . .

Water is essential to life not just because plants and animals all need it, but for another, even more vital reason. There could be no soil without water's unique feature of expanding when frozen; no other known liquid does this. The effect throughout millions of years of water droplets freezing inside rocks was to split them into tiny fragments to make soil. And the great power of the oceans and rivers has eroded the outside of rocks.

Capillary action

Humankind's first settlements were always close to a source of water that they could use both for drinking and for watering their crops. Plants take the water they need from the soil and any surplus evaporates from the leaves. This process is made possible by capillary action, which is the process by which liquids are drawn up tiny tubes. You can see this working by carrying out a simple experiment. First, find a **large leaf** that has as

You will need—

1 large leaf
1 small jar
water
food coloring

much white and yellow on it as possible—do not forget to ask permission before you pick any leaves. Stand it in a small **jar** of **water** and add a few drops of **food coloring**. Leave this for a few days and then you will see that the colored water has been drawn up all through the leaf. You will actually see the leaf's vein structure outlined in color.

Did you know . . .

The world contains an enormous amount of water–about 326 million cubic miles (1.4 billion cubic kilometers). And yet the problems of water supply grow ever more difficult around the world, even though there is as much water circulating around the earth as ever, being constantly recycled by the weather. This is because 97 percent of the world's water is in the oceans, and another 2 percent lies locked up on the polar ice-caps. More than half the remaining 1 percent is underground, often in very deep artesian wells. The remaining fraction includes all water in clouds, rivers, lakes, vapor, reservoirs and snow–including all the water inside every human being (over half our body weight is made up of water). Water is essential to life–whether plant or animal–because it easily dissolves substances necessary for life, for example, oxygen in the blood.

Here are more experiments with water you can try at home. This next one will be very useful if you are going away, leaving your houseplants unattended. You will need two watertight **containers**, a piece of thick **cotton twine** or **woolen yarn**, and **water**.

Procedure

1. First, put the water in one of the containers. Stand both containers close to each other with the top of the empty one at the level of the bottom of the full one.
2. Next, join the two with the cotton or wool, making sure it touches the bottom of both containers. Gradually the lower container will fill and the upper container will empty.
3. To water your plants automatically, use a piece of cotton or woolen material, and a bowl big enough to hold your plants.
4. Stand the plants on the material in the bowl and put the other end of the material in a shallow dish of water. The dish must be at a higher level than the plant bowl. The plants will draw water as they need it.

Siphons

A similar entertainment for your friends is making a siphon. You will need **two glasses**, a length of **plastic** or **rubber tubing**, and **water**—and if you put food coloring into the water you will make it more effective. All you need to do is put some colored water into each glass and cover one end of the tubing with your finger. Fill the tube with water and cover the other end as well. Now put one end of the tube into each glass, remove your fingers and lift one glass slightly higher than the other. You will see that the water immediately starts to flow from the higher to the lower glass. It will keep flowing until the water surfaces are level. If you keep moving first one then the other higher, you can keep the water moving for as long as you like.

Water pressure

For more entertainment using **colored water**, find a **large plastic drink container** and, using a sharp **nail** or **sewing needle**, make a line of holes from the top to the bottom. Ask your friends which hole they think the water will come out of with the most force. Many people will think it will be the top one, but they will be wrong. The weight of water presses down and the water at the bottom spurts out farthest. While you are filling the container, you may find it easier to fill it from a **pitcher**, holding the container at an angle with the holes facing up, until it is nearly full. Alternatively, you could cover the holes with **adhesive tape**, removing the tape as part of the performance.

You will need—

1 large plastic drinks
  container
colored water
nail or sewing needle
pitcher
adhesive tape

Take care!

When making holes in the bottle, make sure you do not push the nail or needle straight through the bottle—you could hurt your hand and also spoil the effect.

Clouds

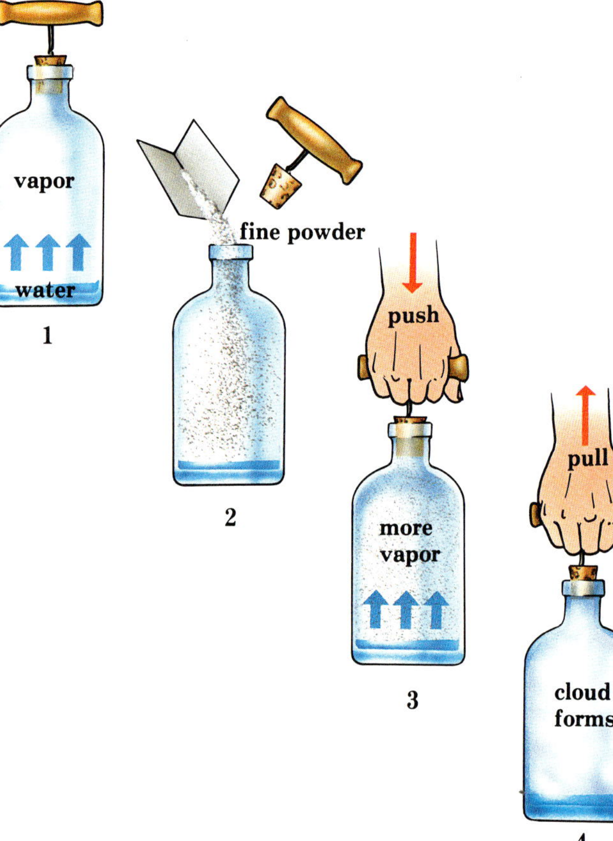

Clouds form in the sky when invisible water vapor condenses to form millions of tiny water droplets. Show how this happens by making miniature clouds in a glass jar.

Procedure

1. You will need a large clear **glass jar** with a pliable, tight-fitting **cork**, and a **corkscrew**. If the cork is dry and brittle, put it in water until it becomes soft.

2. Pour a little **cold water** into the jar and cork it. Some of the water will evaporate to form an invisible vapor (figure 1).

3. After a few minutes, remove the cork. Quickly pour in some dry **talcum powder** or powdered chalk (figure 2). Then replace the cork.

4. Push the cork farther into the neck of the jar, using the corkscrew as a handle. The pressure in the jar increases and causes a sudden rise in temperature. Warm air can hold more vapor than cool air can, so more water evaporates (figure 3).

5. After about 30 seconds, pull the cork out a little way (figure 4). The resulting pressure drop will cool the air in the jar. It can no longer hold so much water vapor. So some of it condenses on the finer powder particles, which are still in the air. The result is a miniature cloud of water droplets.

ADVANCED EXPERIMENTS

In the following pages, there are nine more experiments that demonstrate the laws of physics. These experiments are a little more complicated, using more unusual equipment and often needing help or supervision from an adult.
However, you will enjoy doing all the experiments: learn about fuses, light bulbs, solar energy and pulleys, demonstrate Archimedes' principle, and make your own spectacular siphon fountain.

Layers of liquids

You probably do not think about it, but, like all objects, some liquids weigh more than others. Scientists would say that some liquids are less "dense" than others. For example, oil is less dense (and hence lighter) than water. That is why you often see oil floating on puddles.

You can experiment with many fluids of different densities to find out which ones are lighter than others.

Procedure

1. To do this experiment, you will need a **tall jar**, a **long narrow funnel** (you can make one out of thick paper), **syrup** or **molasses**, ¼ **cup of water** and some **food coloring**, ¼ **cup of kerosene**, ¼ **cup of denatured alcohol**.

2. If you need to make a funnel, use a compass to mark out three six-inch (15 centimeter) circles on a sheet of **thick paper**. Cut out the circles, and cut each one across to its center. Fold each circle around to form a funnel, and fasten it securely with adhesive tape. Cut off the bottom of each funnel with scissors.

3. Carefully pour enough syrup or molasses into the jar to a depth of about ⅜ inch (1 centimeter). Do not let it drip down the side of the jar or the experiment will not work as well.

4. Mix four or five drops of food coloring into the cup of water. Place one of the funnels in the jar, angling the spout end so that it rests against one side of the jar.

Slowly pour in the colored water—go slowly so that it does not stir up the syrup on the bottom.

5. Using the second funnel, add the kerosene in the same way—remember to pour slowly. Then add the denatured alcohol the same way.

6. If you have added all the liquids carefully, they will not mix. Instead they will float in separate bands because they have different densities. The lighter the liquid, the higher up it floats.

7. To prove this, do the experiment again, but this time pour in the kerosene before the water. Leave the jar for a while to allow them to settle, and you should find that the water settles out below the kerosene even though it was added after the kerosene.

8. You can also experiment to see whether a small solid object is heavier or lighter than some of the liquids in the jar. Cut a slice of **candle**, then gently lower it onto the surface of the liquids.

9. You will find that the candle sinks through the layers of denatured alcohol and kerosene to float on top of the layer of colored water. This means that the candle is heavier than denatured alcohol and kerosene, but lighter than water or syrup.

10. Try a similar test with other small objects, such as a **marble**, to compare their densities with those of the liquids.

Kerosene and denatured alcohol are poisonous, so keep them well out of the reach of small children and animals.

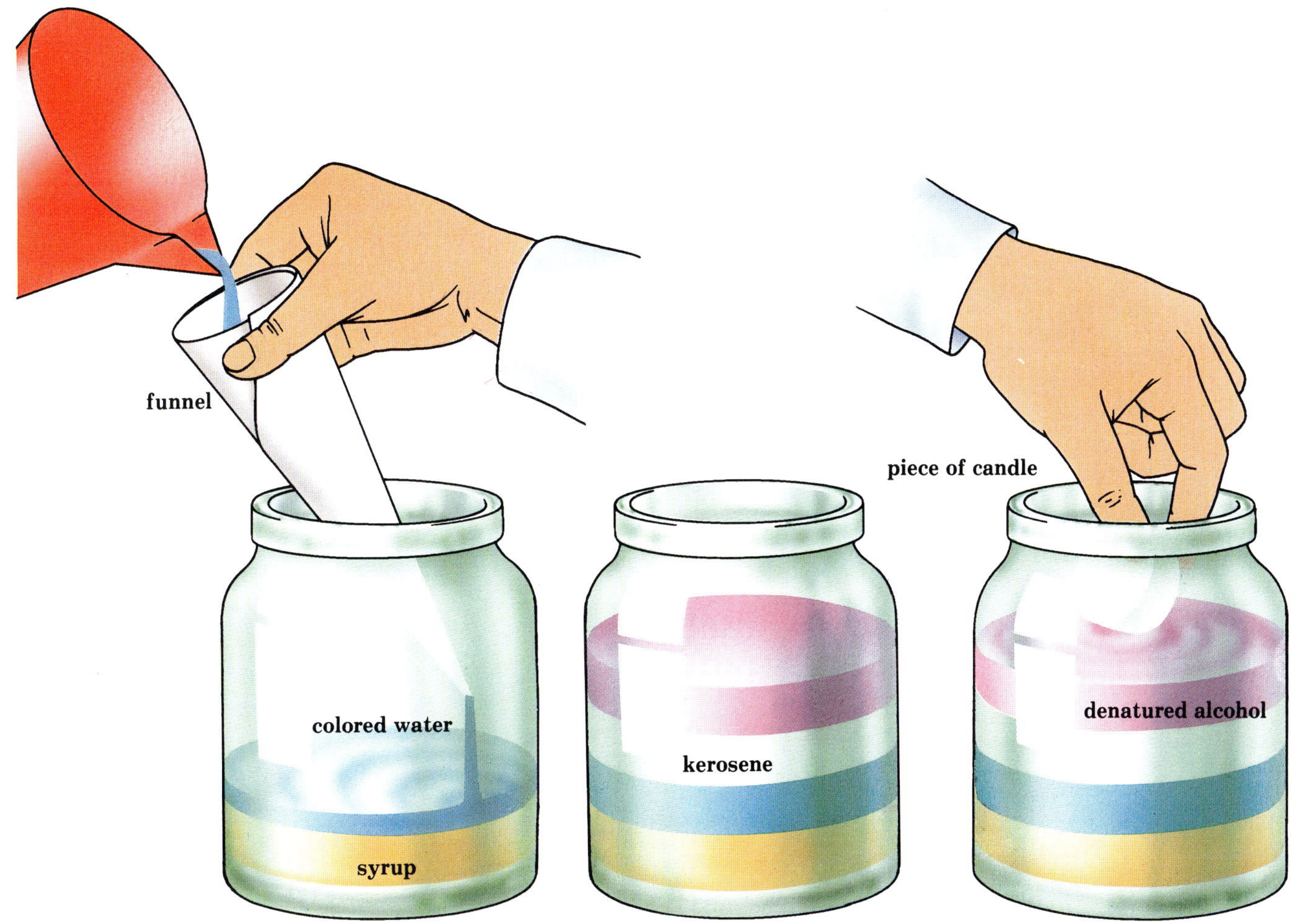

Siphon fountain

This experiment can be rather messy, so wear old clothes and do it outside where spilled water will not matter. You will need an empty, clear plastic bottle at least 8 inches (20 centimeters) high and a cork or rubber stopper which fits it tightly. Drill two holes through the cork into which plastic drinking straws can be tightly fitted. One straw should project through the cork into the bottle by one inch (2.5 centimeters), the other by six inches (15 centimeters). If necessary, use glue to insure an airtight seal around the straws. You will also need two large jars, one empty and the other half full of water. The jar containing water should be on a box or step between six and twelve inches (15 and 30 centimeters) above the empty jar.

Procedure
1. Fill the **bottle** completely with **water**. Push the **cork** with **straws** attached firmly into the top of the bottle.
2. Seal the ends of the two straws using **modeling clay** or **paper clips**.
3. Carefully turn the bottle upside down so that the straw that projects furthest into the bottle is over the jar that is half full of water.
4. Remove the seal from the straw over the half-filled jar and immediately lower the end of the straw under the surface of the water. The water level in the jar should rise while that in the bottle falls. In other words, water flows from the bottle to the jar. After a few seconds this flow of water should stop.
5. If the tips of both straws are now showing above the water surface in the bottle, you either have a leak or are using a bottle that is too small. In either case, you will need to start again. If neither straw shows above the water surface in the bottle you will need to introduce a little air into the bottle by lifting the straw out of the beaker of water momentarily.
6. You can now remove the seal from the straw over the empty jar. Water will flow quite rapidly from the full jar up into the bottle, forming a fountain, and then down into the empty jar.

It may seem strange that water should, of its own accord, travel up a drinking straw to form a fountain but this can be explained by air pressure. With the bottle fountain, when you released the seal from the first drinking straw and placed it over the beaker of water, gravity pushed the water downward out of the bottle. As water drained from the bottle, the volume of air inside the bottle increased but the actual amount of air remained the same, since no air could get into the bottle from outside. When the volume of a gas increases in this way, its pressure decreases. The pressure of the air in the bottle will no longer compensate exactly for the external atmospheric pressure. The flow of water stops when the pressure difference exactly matches the force of gravity on the water column.

When the seal on the second straw was released, the pressure within the bottle was not enough to prevent gravity from pulling a column of water through the straw because this column of water was longer, and hence heavier than that on the other side. As water flows out of the bottle, the air pressure within the bottle is reduced further so that it is not enough to prevent external atmospheric pressure from pushing water up the straw from the full beaker into the bottle, therefore forming the fountain.

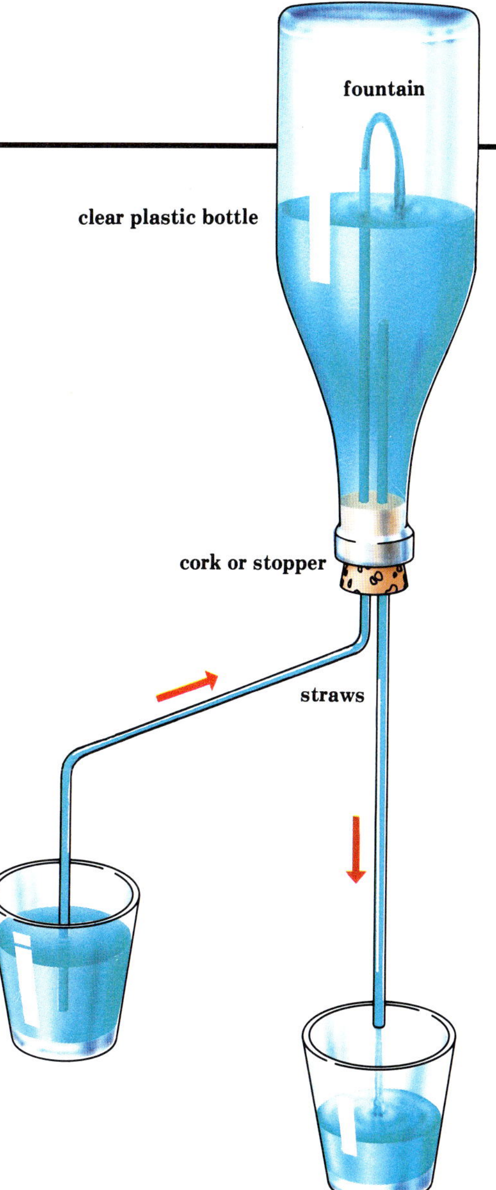

Solar energy

A solar cell is an electronic device that converts light energy into electricity. This project shows how the current from a solar cell can be greatly increased by using a reflector to concentrate the sun's rays.

Procedure
1. Purchase a **solar cell** from an electronics dealer. An **electrical meter** is used to show the current produced by the cell. For this, use any **galvanometer** designed to measure direct (one-way) electrical current in the range 0 to 10 milliamps. (One milliamp is equal to 1/1000 of an amp.)
2. **Glue** the back of the solar cell to the side of a **cardboard box**. This is used as a support.
3. With two pieces of **insulated wire**, connect the two wires of the cell to the ends of the galvanometer coil.
4. Shield the cell from light. Then turn the galvanometer until its compass needle points to the coil. This will occur when the coil lies in a north–south direction.
5. Expose the cell to the sun's rays. The needle should move slightly to one side.
6. Now use a concave mirror (the kind sometimes used for shaving) to reflect the rays onto the cell. The mirror concentrates the light rays. Position the mirror to obtain a small but bright patch of light. The galvanometer needle should now swing much farther to the side. This indicates that a much stronger current is now flowing. The mirror collects much more energy from the sun than can be captured by the solar cell by itself.

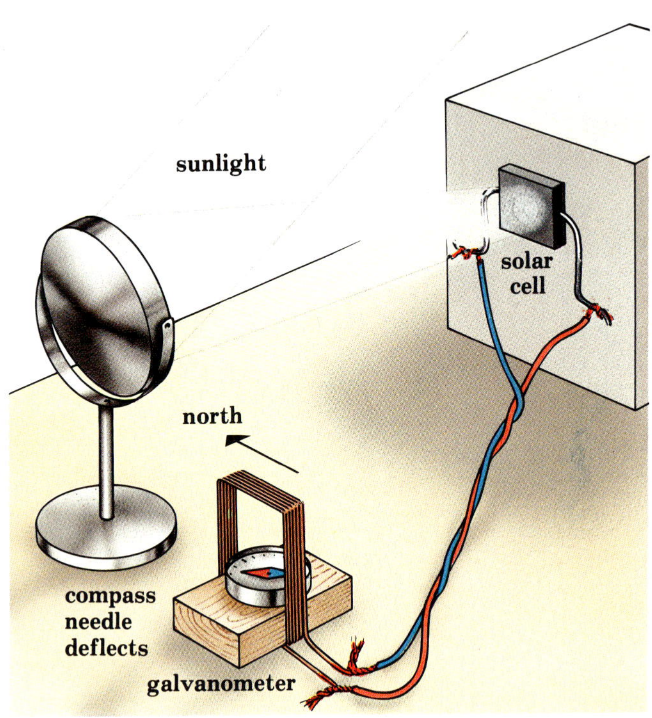

Air thermometer

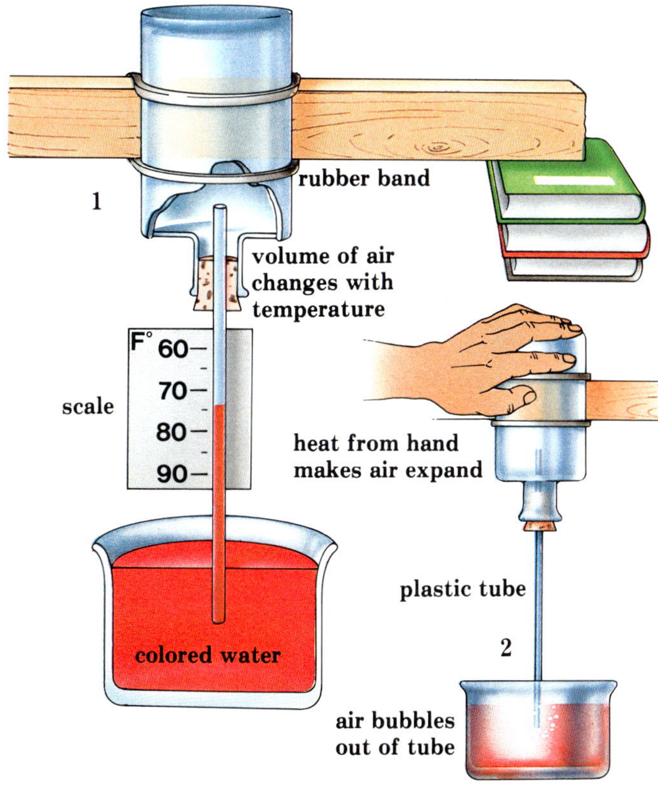

Most thermometers work by measuring the expansion of a liquid—alcohol or mercury. But the thermometer shown here uses the expansion of air.

Procedure

1. You will need a **small bottle** with a **tight-fitting cork** and a short piece of **thin, transparent tube**. The tube can be made of glass or plastic. Some drinking straws are suitable for this purpose.
2. Make a neat hole through the cork, just big enough to take the tube.
3. Push one end of the tube into the hole and push the cork into the neck of the bottle.
4. Pour **water** into a **jar** and add a little **ink** or food coloring.
5. Place the jar on a table and support the bottle above it so that the tube dips into the colored water. One way of supporting the bottle is shown in figure 1.
6. **Glue** a piece of **white cardboard** onto the tube.
7. Warm the bottle with your hand to make the air expand. Some of it will bubble out through the water (figure 2). Now remove your hand and allow the bottle to cool down. The colored water will rise up the tube.
8. Put the air thermometer in a cold room and switch on a heater. As the room warms up, the liquid will move down the tube. Take readings from an ordinary **thermometer** and transfer them to the card to make a scale for the air thermometer (figure 1).

You will need–

1 small bottle
1 tight-fitting cork
short piece of thin
    transparent tube
ink or food coloring
glue
white cardboard
1 thermometer
books for support
rubber bands

Archimedes' principle

Archimedes was a Greek scientist who discovered that "a body partly immersed in a fluid is buoyed up with a force equal to the weight of the fluid displaced by the body". This force is what makes objects float, and is the principle behind all boats and even modern oil rigs that are supported on floating pontoons. To investigate Archimedes' Principle you can do an experiment using a spring scale, a large water container, and a block of wood.

Procedure

1. You will need an object that will float—a **block of wood** is ideal. (If the object sinks and touches the bottom of the water container, the object will be supported by the base as well as the buoying forces of the water, and the experiment will not work.)
2. Drive a small **nail** into the block of wood so that it can be weighed later on in the experiment.
3. Hang a **fisherman's scale** from a secure hook, preferably over a **sink** or a large **drip tray**.
4. Find a container, such as an **ice cream carton**, that is large enough to hold the wooden block.
5. Tie a piece of string securely to the container, and hang it from the fisherman's scale. Fill the container with water right up to the brim—do not worry if some water spills out.
6. At this point, make a note of the combined weight of the container and the water.
7. Now carefully float the block of wood in the water. Some water will spill out of the container. Remember that the weight of the block is the force with which the block of wood presses down on the surface that supports it. The force is caused by gravity attracting the block toward the center of the earth. When the block floats in water, the force from the block pressing down into the water must be equal to the buoying force. If this were not true, the block would very soon sink.
8. If the buoying force is equal to the weight of the water displaced, as Archimedes predicted, the weight registered on the scale will not change.
9. Remove the wooden block from the water, being careful not to spill any water from the container. The weight registered on your scale should now change.
10. Note the new reading of the water and container, and subtract it from the original one—this will give you the weight of the block of wood.
11. As a further check that Archimedes was right, take the container of water off the scale, tie a piece of string to the nail in the wooden block, and weigh the block. The weight will be the same as that arrived at after subtraction in step 10.
12. This experiment will work with any fluid (from a soft drink to mercury) and any floating object that can be supported by the fluid.

Take care!

Use a small hammer to drive the nail into the block of wood, but be careful not to hit your fingers.

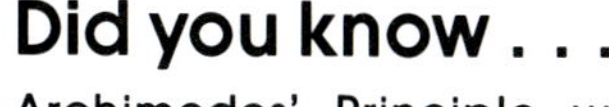

fisherman's scale

Pulleys

You will need—

1 length of coat hanger wire
4 empty spools
1 screw hook
1 weight
1 length of cord
fisherman's scales

Pulleys are simple machines for lifting heavy weights or for applying large forces. Pulley systems are used, for instance, in automobile repair shops for lifting engines, and also on sailing ships. The ropes fastened to the sails are on pulleys so that the sailors can raise and lower the heavy sails against the force of the wind.

Simple pulleys

You can make a simple pulley by cutting a **length of wire** from a metal coat hanger and bending the ends round to make a triangular shape. Slide an empty **spool** between the cut ends of the wire on one side of the triangle and push the ends together so that the spool cannot fall out. Now fasten a **screw hook** into a beam or the top of a doorway and hang the point of the triangle over the hook.

If you now run a length of **cord** over the spool and tie one end to a **weight** such as a bucket with stones in it, you will find that the bucket is lifted from the floor when you pull on the free end of the cord. However, a pulley arrangement like this will not help you to lift weights that would normally be too heavy to lift by hand—in fact you will have to use slightly more force than would be needed to lift the bucket from the ground by its handle. This is because there is some friction

between the spool and the wire triangle, and a little force is needed to overcome this friction.

The way to make a pulley which will help you to lift heavy weights is to add another spool to the system. Make another triangle from coat hanger wire and slip it around the handle of the bucket before you put the spool over the cut ends. Now tie the end of the cord to the wire of the top triangle, loop it under the lower spool and over the upper one. If you now pull on the cord, it will take less effort to raise the bucket, but for every inch that the bucket rises, you will need to pull the cord two inches.

In theory, the effort needed to pull the cord will be only half the weight of the bucket, but the friction in the pulleys will make the true figure a little more than half. You can check this: first, weigh the bucket with a fisherman's scale. Then tie the hook of the scale to the free end of the cord and pull on the other end of the scale. The scale will read just over half the weight of the bucket when the bucket starts to lift off the floor.

If you build a set of pulleys with four spools, the effort needed will be even less, but for every inch that you lift the bucket, you will need to pull the cord through four inches. As before, check the effort needed to lift the bucket against the bucket's original weight.

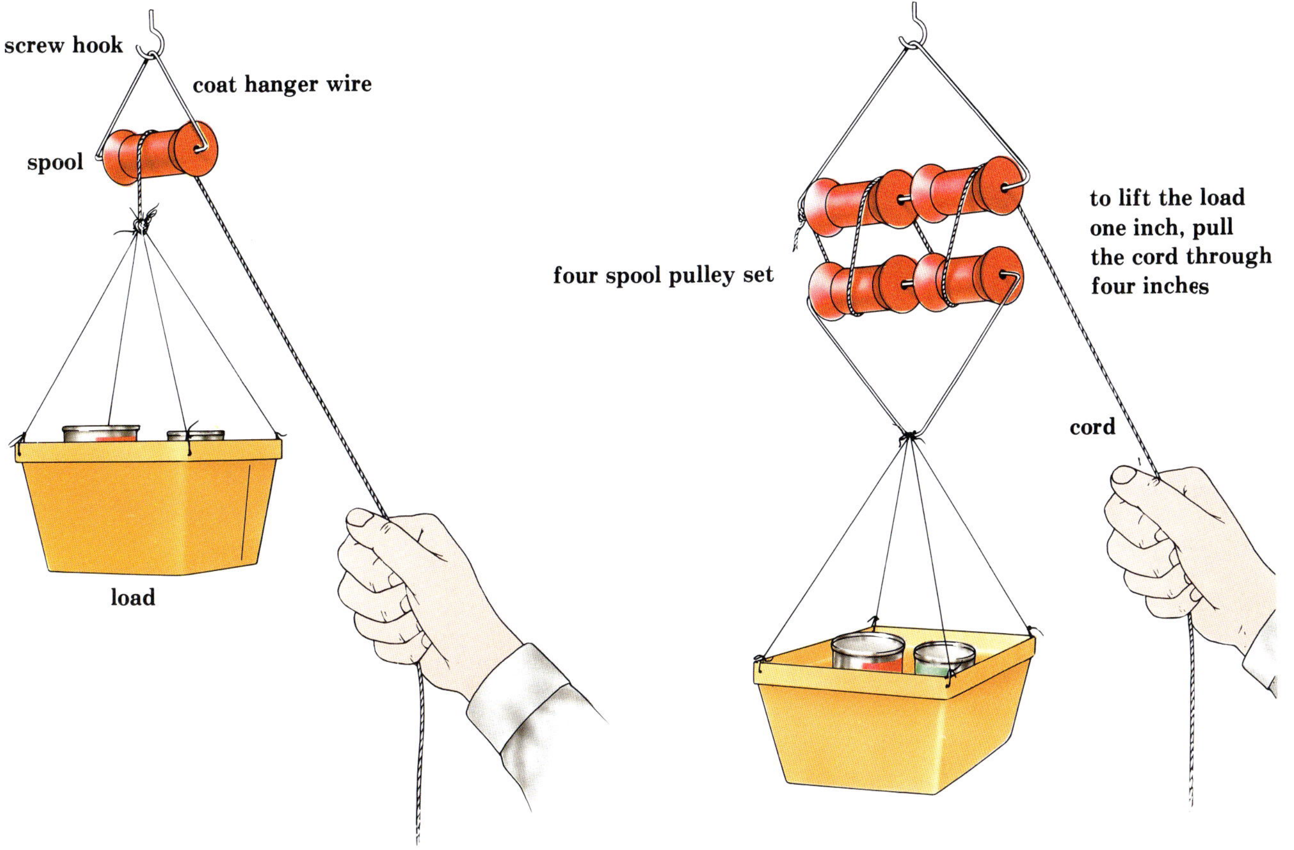

Take care!

Use a firm step ladder when fastening the screw hook into the beam or doorway. And make sure you remember to ask permission before you do it.

Fire alarm ⌗

1 empty soft-drink can
1 magnet
1 old pair of scissors
1 file
1 pair of pliers
wire
1 block of wood
1 wooden board
1 tack
glue
1 small electric buzzer
1 battery
screws
1 screwdriver

Demonstrate the action of a bimetallic strip by making this model fire alarm. The special strip is made from two metals that expand at different rates when heated. This causes the strip to bend, so that it completes a circuit to sound a buzzer.

Procedure

1. For the bimetallic strip, use a strip of steel and a strip of aluminum. These can be cut from an empty **soft-drink can**. The tops are made of aluminum for ease of opening, and the rest of the can is often steel, as this is cheaper than aluminum.

Use a **magnet** to check that you have a suitable can. The magnet should cling to the steel body of the can, but if the top is made of aluminum, the magnet will not cling to it.

2. Use an old pair of **scissors** to cut the strips of metal. This should be fairly easy, as soft-drink cans are usually quite thin. But take care that you do not cut your fingers on the metal. The two strips should be the same size and at least 2 inches (5 centimeters) long. Use a **file** to smooth over any sharp edges.

3. Scrape clean one end of each strip using the file. (The metal has a protective coating which must be removed to obtain a good electrical connection.)

4. Hold the two strips of metal together so that the scraped ends are in contact. Then use a pair of **pliers** to fold the ends over. The steel strip should be on the inside of the fold. Fold the ends over again so that the two pieces of metal are now firmly clamped together (see figure 1).

5. An electrical connection must be made at the other end of the bimetallic strip, so scrape the aluminum side clean. Then fold this end of the strip over so that the end of a piece of **wire** is gripped firmly by the aluminum (figure 1).

6. Cut out another strip of steel from the can. This one will be used to make contact with the bimetallic strip. Shape one end of the strip to a point and bend it as shown (figure 2). Clean a patch near the other end for a connection to be made. And clean a patch on the steel side of the bimetallic strip for the point to make contact.

7. The bimetallic strip and the steel contact strip must be mounted so that they almost touch. This is done by tacking them to a **block of wood** (see illustration). The **tack** through the contact strip also serves to clamp a connecting wire.

8. **Glue** the block with the metal strips attached to a **wooden board**.

9. Obtain a small **electric buzzer** and a suitable **battery** for it. Screw the buzzer to the board and then complete the fire alarm by making the connections shown in the illustration.

10. Test the circuit by pushing the steel point against the bimetallic strip. This should complete the circuit and make the buzzer sound.

11. Adjust the contact strip by bending it so that the point almost touches the bimetallic strip.

12. Demonstrate the alarm by placing it near a heater. The aluminum will expand more than the steel in the bimetallic strip. This will make the strip bend and touch the steel point. With the circuit thus completed, the buzzer will sound.

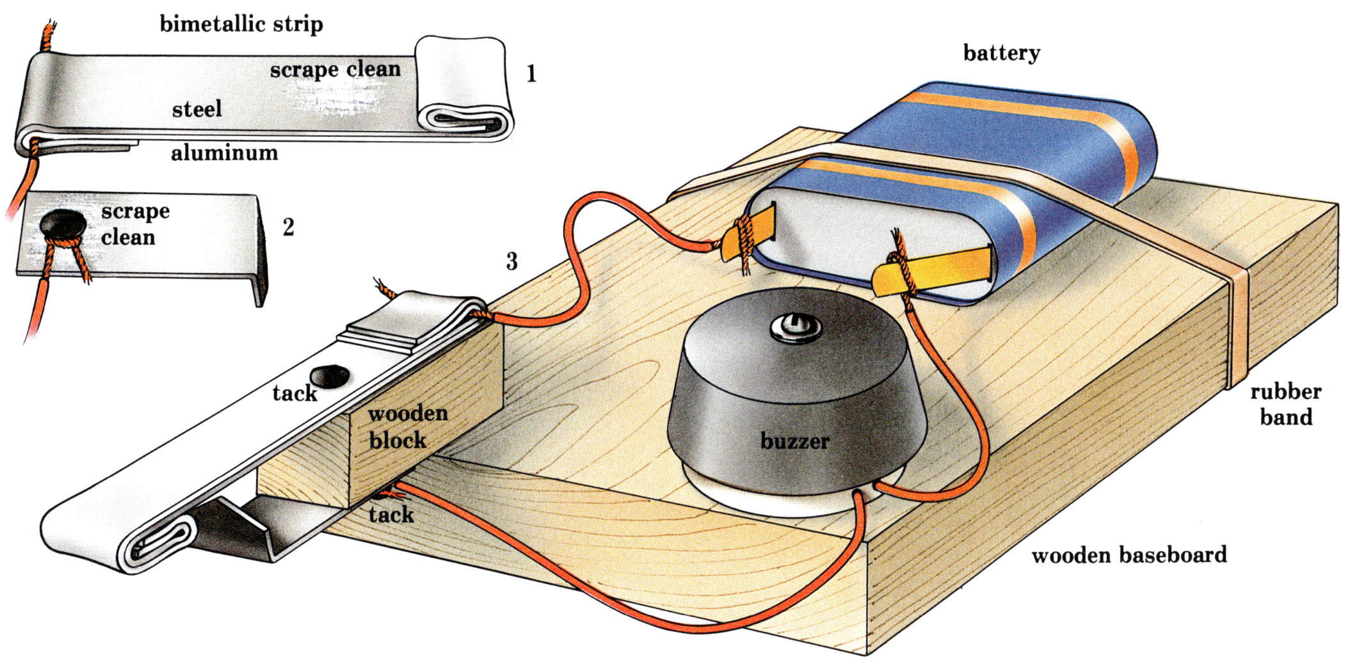

Take care!

Wear thick gloves to protect your hands when cutting and filing the soft-drink can. It is important to file the edges of the metal smooth so that there are no sharp edges to cut your hands during the experiment.

Fuses ⊕

You will need—

1 9-volt battery
1 small flashlight bulb in a holder
copper wire
steel wool
1 small can or lid
1 switch (optional)
1 small block of wood
2 thumbtacks
1 paper clip

Electricity is a wonderful thing. Without it, modern life would be very different and very much harder, with none of the household appliances such as washing machines, microwave ovens and food processors to take the drudgery out of housework. In common with most powerful forces though, electricity must be handled with care or it can kill a person or set buildings on fire. To keep electricity from damaging people or buildings, electrical wiring has to be well insulated with a plastic material that does not conduct electricity. If there were no insulation, bare wires could touch each other and so create a short circuit. It is this sort of short circuit which leads to large amounts of electrical energy being discharged. The result may be a fire or the electricity may go through a person, causing terrible burns and shock, even death. To keep this from happening, every circuit that carries a heavy load of electricity must have a weak point in it. If there is too much electricity passing through, the weak point heats up, melts and falls away, breaking the circuit. We call this weak link the fuse. The fuse is still the most common way of protecting an electrical circuit from overheating, although some modern installations have sophisticated electronic circuit breakers which measure the resistance to the passage of electricity down the wire. If they detect any danger, they switch the whole circuit off immediately. To demonstrate how a conventional fuse works you will need a 9-volt **battery**, a small **flashlight bulb in a holder, copper wire, steel wool**, a small **can** or **lid**, a **block of wood**, two **thumbtacks** and a **paper clip**.

Procedure
1. First make the switch: place the two thumbtacks about 2 inches apart on the block of wood, and attach the twisted paper clip to one of the thumbtacks.
2. Connect the bulb to one terminal of the battery with copper wire.
3. Use one strand of steel wool to connect the bulb to another strand of copper wire.
4. Connect this wire to one thumbtack on the switch, and connect the other thumbtack to the battery with more wire.
5. Place the can or lid under the strand of steel wool to catch any hot bits that fall.
6. Now, when you complete the circuit by pressing the paper clip against the thumbtack (switching on), the light will go on for a short time. The steel wool fuse will soon start to glow and will quickly melt, breaking the circuit and making the light go out.

In an average house, you will usually find three types of circuit: a lighting circuit which runs the ceiling and wall lights; a more powerful circuit for small appliances which need more electricity; and an even more powerful circuit for large appliances like stoves and showers. WARNING: if you ever need to mend a fuse at home, make sure you fit the correct fuse. Do NOT use wire.

Did you know . . .
Not all fuses are electric; fuses are also used to set off explosions in blasting or quarrying, which rely on a trail of black gunpowder. When lit, this burns slowly toward the explosive.

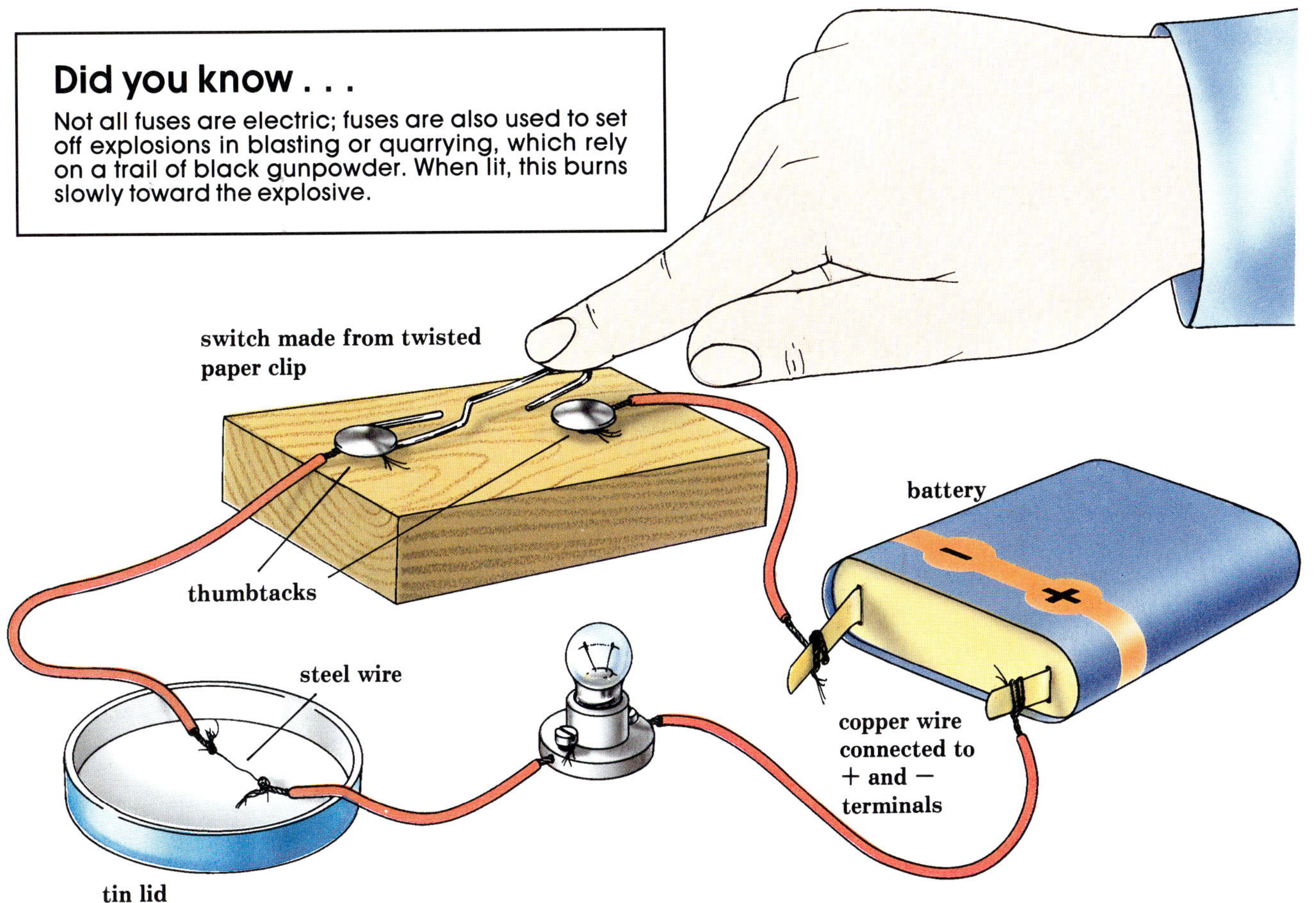
switch made from twisted paper clip
thumbtacks
steel wire
tin lid
battery
copper wire connected to + and − terminals

Light bulb

1 small jar
thin iron wire
2 long nails
copper wire
1 9-volt battery
1 switch (optional)
cardboard or 1 plastic lid

Like a fuse, a light bulb uses the principle of resistance to electrical current. This resistance produces heat and light but, unlike the fuse which glows and melts, the wire in a light bulb (called the filament), which is often made of wolfram (tungsten), glows and does not melt. The glass globe of a light bulb contains a partial vacuum. Most of the air is drawn out and minute traces of special gases such as argon and nitrogen are put in. The first practical light bulbs were invented in the latter part of the 19th century by Thomas Alva Edison. He used thin, carbonized bamboo as the filament. At the same time, another inventor named Joseph Swan was trying a similar experiment using cotton thread as a filament. Both Edison and Swan were following in the footsteps of an American engineer named Starr who invented the first carbon-filament incandescent lamp as early as 1841. Starr's light did not last long before it burned out.

Making a light

These first carbon-filament lights were popular because they were easier to use than gas lights or oil lamps. They did not smell or make a mess but they were expensive and did not give out a great deal of light. However, they were widely used until around 1908 when the tungsten filament became usual. It is possible to make a crude light bulb at home but, like the early inventors, you will not be able to light very much with it nor will it stay lit for long. You will need a small **jar**, some thin **iron wire**, two long, clean and non-greasy **nails**, **copper wire**, a **9-volt battery** (a **switch** is useful if you have one), and some **cardboard** or a **plastic lid** for the jar.

Procedure

1. First, make two holes in the lid and push the nails through, leaving the heads of the nails sticking up about ¼-inch (6 millimeters).
2. Wrap the iron wire around the pointed ends of the nails—this is your filament.
3. Place this assembly in the jar—make sure the nails and wire do not touch the jar.
4. Next, attach one piece of copper wire to each nail head. If you have a switch, link one piece of wire to the switch and connect the switch to a battery terminal with more wire.
5. Attach the other wire direct to the other battery terminal.
6. When the switch is turned on, the "lamp" should glow. If it does not, try putting an extra battery into the circuit. This should be connected up "in series." In other words, the copper wire from one side of the bulb should go to the positive terminal of one battery. The other wire from the bulb goes to the negative terminal of the second battery, and a third piece of wire should be used to join the negative terminal of the first battery to the positive terminal of the second battery. This arrangement increases the voltage and should light the lamp without difficulty. If you do not have a switch, simply complete the circuit by touching the wire to the terminal of the battery.

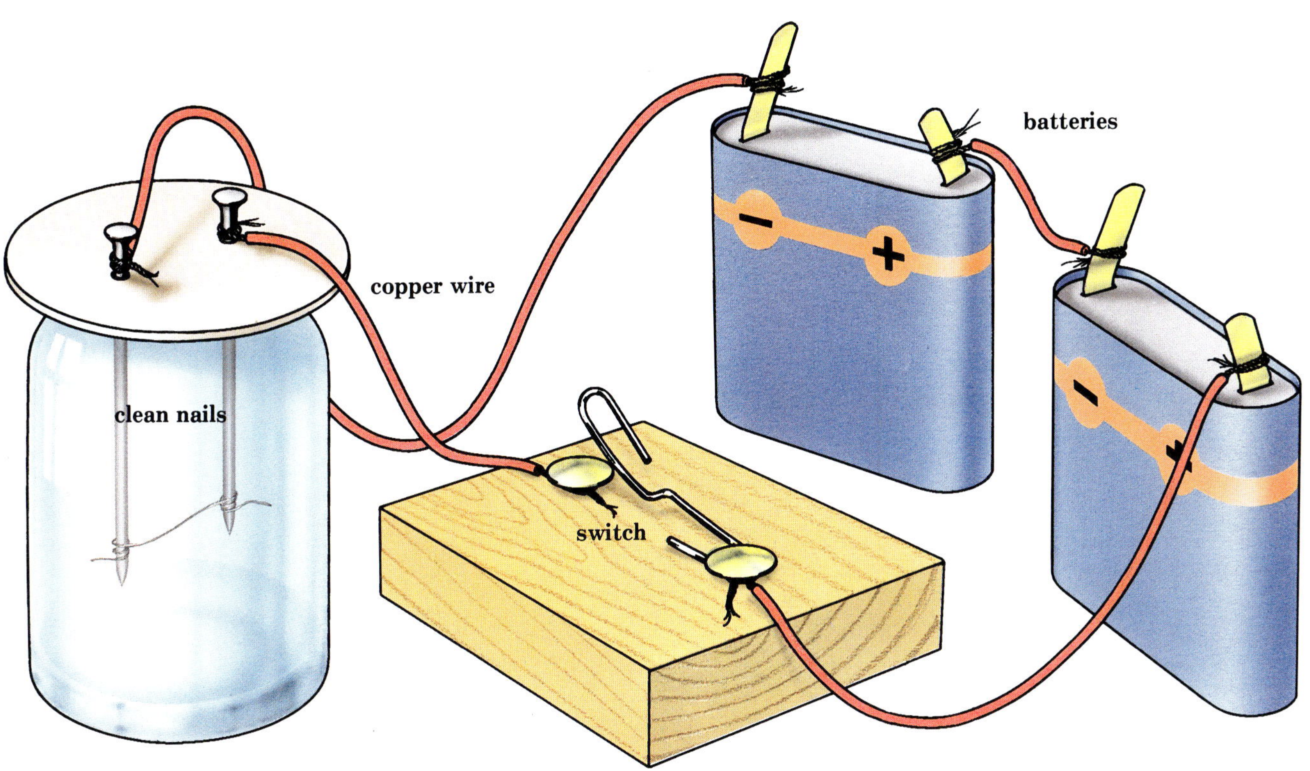

Take care!

Make the holes in the jar lid using a hammer and a nail the same size as the two nails used in step 1, keeping your fingers well away from the nail head as you hammer.

If you do not have a switch, see page 36 for instructions to make the simple switch shown above.

Safety first!

The safe and simple projects and experiments in this book cover many fields of science and technology, and have been designed to demonstrate and explain important scientific principles in an interesting and straightforward way. Good scientists take care to protect themselves and other people, so always follow these rules for perfect safety.

***Fire** Take care when using matches or candles, and keep a pail of water nearby in case of accidents.

***Main line electricity** Use low power batteries as directed in this book. Do not use the main line—it can kill.

***Chemicals** Use with care, label clearly, and store out of reach of young children and animals.

***Sharp edges** Where possible, file edges smooth, and always protect your hands with thick gloves.

And remember . . . before you begin, always get permission from an adult, and if in doubt, ask for help!

Safety

The chemicals used in the projects can all be handled safely. Most are common household substances such as salt, baking powder, vinegar, etc.

When a project calls for an electricity supply, there is no danger of electric shock because a low-voltage battery is used.

A few projects involve the use of a flame or heat from a stove. In such cases, younger experimenters should be supervised by an adult, but the procedure is as safe as cooking.

Supervision

The projects have been graded according to the need for adult supervision. Where a project is marked with an ✧, it means that, for complete safety, an adult should assist the young experimenter with some aspects of the project. In many cases, the adult's assistance will be limited to helping with some parts of the project, such as using a hammer and nails, and then letting the experimenter continue the project with only background supervision. Similarly, it may be necessary for an adult to handle matches or candles.

In other cases, the adult can prepare the materials which are needed for the experiment. For example, if the project includes accurate cutting out with a sharp knife, the experimenter need not handle the knife if the adult does the cutting out beforehand.

It is a good idea for an adult to be present when the project involves breakable objects like glass jars. A little guidance will minimize the risk of breakages.

Sharp edges

Wherever it is possible, the materials chosen for the projects are the safest ones that can be used. Sometimes, there is a choice of materials. For instance, the risk of sharp edges is reduced if you use plastic glass, but if you do use conventional glass, the chance of getting cut will be minimized if you are careful. First, protect your hands with gloves and use a small file or some sandpaper to remove any sharp edges from the glass. Handle the glass carefully and never put too much strain on it.

Similar precautions should be taken if the project involves cutting metal. Again, wear gloves and file off any sharp points. For extra safety, cover the cut edges with thick adhesive tape. If the project uses an empty tin can, try to find a can with a push-fit lid. This means you do not need to use a can opener, which will leave a sharp edge.

Chemicals

The chemicals used in the projects are all harmless, but they should still be treated with care. Keep each chemical in a labeled jar, and make sure that the jars are stored out of reach of inquisitive small children. Do not experiment with chemicals anywhere near food. Cover the worktable with old newspaper; it will catch any spilt chemicals and can be thrown away later. After the experiment, wash thoroughly the jars or dishes that have been used and pour the old chemicals down the sink, flushing them away with lots of water. Finally, wash your hands to remove any traces of chemicals on the skin.

Fire

You should take extra care with those projects which involve matches or candles. Organize the worktable so that there are no scraps of paper lying around and make sure that you are not wearing loose clothing, such as a tie or scarf which might accidentally catch fire. Check that you have all the materials needed for the experiment before you begin and arrange them on the worktable so that you do not need to reach over the candle. Always keep a pail of water close at hand just in case there is an accident.

Procedure

Before starting work on a project, it is important to read the instructions through to the end and to form a clear idea of what has to be done, and in what order. Materials and tools needed are listed in the margin and are spelled out in **bold type** when they are first mentioned. Make sure that everything is at hand when it is needed. Many of the projects or experiments can be carried out more smoothly if a little preparatory work, like weighing or cutting out, is done beforehand.

Science and the future

There are more and more opportunities for scientists in the modern world. Every year, new scientific discoveries help to change the world we live in. Most aspects of our life, including transport, entertainment, medicine and industry are changing rapidly because of the new inventions and discoveries that scientists are making. When you work on the projects in this book, you will learn many of the basic scientific principles which have helped important scientists to make their contribution to our world. Maybe one day, if you decide to become a scientist, you will join the great men and women of science who will create the world of the future. Who knows what you could achieve?

Words you need to know

In this book, you may find some words that you haven't seen before. These four pages explain as simply as possible what these words mean and will help you to understand exactly how to do the projects or experiments. Some words are special descriptions invented by scientists, and so are often very complicated to explain—in fact, whole books have been written about them! Of course, there isn't space in this book to include these very long explanations, but if you want to read more about any of them, ask your teacher or librarian to help you find a book.

Air pressure
The constant pressure of air on all objects. At sea level, about 14.7 pounds per square inch (1.03 kilograms per square centimeter)

Aluminum
Lightweight, silver-colored rust-resistant metal

Amp(ere)
The standard unit for measuring electricity

Argon
A colorless, inert gas making up one percent of the atmosphere and used in light bulbs

Astronomer
A scientist who studies the stars and planets (an astrologer believes the stars effect human actions)

Atmospheric pressure
Identical to air pressure

Bimetallic
Made of or using two metals; two bands of metals with different heat expansion used to measure or control temperature

Capillary
Hairlike, of minute diameter; in capillary attraction/repulsion, the force by which liquid is drawn up or down in capillary tubes

Carbon
Non-metallic chemical
element found in
diamonds and charcoal

Channel
Natural or artificial hollow
bed or area of liquid; in
broadcasting, band of
transmitting frequencies

Circuit
In electricity, the path of
a current or apparatus
through which it passes

Circuit breakers
Device that interrupts the
flow of electric current
automatically

Compasses/compass
An instrument which can
be used both for taking
measurements and
drawing circles; navigation
instrument showing true or
magnetic north

Compensate
To counterbalance, offset
or make up for an effect

Concave
Curved inward like the
interior of a cave

Condense
To make denser or more
compact; to reduce from
a gas to a liquid

Conduct
To transmit heat or
electricity

Connection
In electricity, a circuit

Denature
To change the nature of; to
render unfit for human use

Deposition
In geology, the laying
down of clay or rocks by
action of wind or water

Effluent
Waste material flowing
as discharge from, for
example, an industrial site

Equator
Imaginary circle around
the earth at an even
distance from the north
and south poles

Galvanometer
Instrument for detecting
and measuring small
electric currents

Globe
Perfectly spherical body

Gravity
Weight, heaviness; force
which attracts bodies to the
center of the earth

Gyrocompass
A compass giving true
north and bearings from it
by means of a gyroscope

Gyroscope
A wheel set in a ring so that its axis can turn in any direction. When spun quickly, it keeps its first plane no matter which way the ring is turned; used in ships or aircraft

Hydrogen
Colorless, odorless gas, the lightest substance known, very inflammable

Inertia
In physics, the tendency of matter to remain at rest or, if moving, to continue in the same direction

Kerosene
Thin oil distilled from petroleum and used widely in lamps and stoves

Lubrication
Fluid or oil used to reduce friction

Mercury
Whitish, normally liquid, metal used in thermometer or barometer

Microscope
Instrument which magnifies objects by means of lens

Microweld
Very small weld

Miniature
Very small, tiny

Molecule
The smallest particle of an element that can exist in the free state and still keep its character

Nitrogen
Colorless, odorless gas, forming four-fifths of the atmosphere

Perpetual
Lasting forever, eternal

Physicist
A scientist, expert in physics

Pontoon
Flat-bottomed boat used as floating bridge

Precession
Slow movement of axis of spinning body around another axis

Principle
The law of nature by which a thing operates

Repel
To push away; in magnetism, the opposite to attract

Scale
Set of marks at measured distances

Silt
Sediment deposited by water in channel or lake

Siphon
Tube used to carry fluid over edge or ridge by exploiting atmospheric pressure

Solar cell
Device used to convert the sun's energy into electricity

Solar energy
Energy from the sun: every forty minutes, it gives the earth's surface as much energy as humanity uses in a year

Solution
Substance dissolved in a liquid, usually water

Sophisticated
Complex, highly developed

Static
Not moving, fixed, the opposite of dynamic

Strand
One of the lengths of a material twisted together to form a string, wire or rope

Surplus
An amount over what is needed, excess, extra

Suspend
To hang something freely to allow movement

Temperature
Degree or level of heat

Tension
State of being stretched or stressed

Terminal
In electricity, one end of an electric circuit

Tungsten
Steel-gray heavy metallic element used to make electric light filaments

Weld
To join pieces of metal by application of great heat or by pressure

Index